BENEATH THE EARTH

—— ✦ The Facts and the Fables ✦ ——

✦

Finn Bevan
Illustrated by Diana Mayo

CHILDREN'S PRESS®
A Division of Grolier Publishing
NEW YORK • LONDON • HONG KONG • SYDNEY
DANBURY, CONNECTICUT

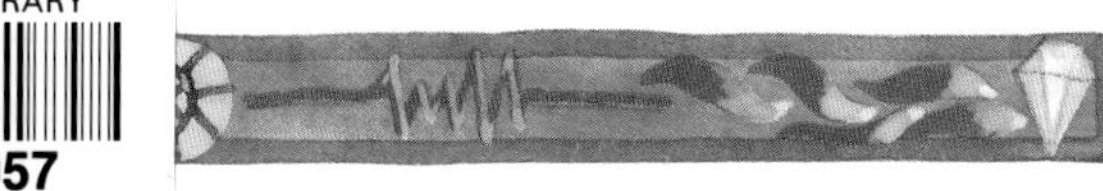

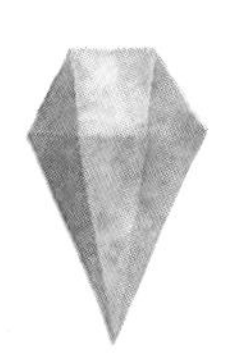

Picture Acknowledgments:
AKG London p. 11 (Erich Lessing); Axiom Photographic Agency p. 18 (Jim Holness); C.M. Dixon p. 10; Werner Forman Archive pp. 6, 15 (British Museum, London); Hutchison Library p. 24b; Planet Earth Pictures pp. 14 (Krafft), 24t Ken Lucas; Slide File p. 20.

Series editor: Rachel Cooke
Art director: Robert Walster
Designer: Mo Choy
Picture research: Sue Mennell

First published in 1998 by Franklin Watts
First American edition 1998 by Children's Press
A Division of Grolier Publishing
90 Sherman Turnpike, Danbury CT 06816

Visit Children's Press on the Internet at:
http://publishing.grolier.com

Bevan, Finn.
Beneath the Earth : the facts and fables / Finn Bevan :
illustrated by Diana Mayo.
p. cm. -- (Landscapes of Legend)
Includes index.
Summary: a collection of myths and stories from various cultures that attempt to explain elements of the underground including earthquakes, volcanoes, caves, and springs.
ISBN 0-516-20954-X (lib.bdg.) 0-516-26302-1 (pbk.)
1, Underground areas--Folklore. 2. Caves--Folklore.
3. Earthquakes--Folklore. 4. Volcanoes--Folklore. [1. Underground areas--Folkore. 2. Caves--Folklore 3. Earthquakes--Folklore.
4. Volcanoes--Folklore.] I. Mayo, Diana, ill. II. Title.
III. Series
GR665.B47 1998
398'.36--dc21

97-51338
CIP
AC

Printed in Singapore

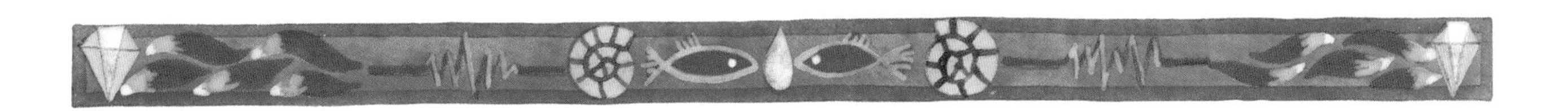

Contents

✦

Mysteries of the Earth

Ancient people could only guess what lay beneath their feet. They knew about caves, which led them to believe that Earth was hollow, crisscrossed by a maze of underground passages — a cold, dark world. Others knew about volcanoes and thought it must be extremely hot underground. But no one knew for certain, so myths and legends grew up to explain the distant, shadowy regions beneath the Earth.

Worlds Underground

Many people believed that the Underworld lay underground: a deep, dark place to which people's souls traveled after death. Like everything else, the Underworld was controlled by the gods. There were many different versions of what the Underworld looked like.

In African legend, Earth was divided into two parts, separated by an ocean. Above was the land of the living, rising up like a high mountain. Below the ocean lay the Underworld, which was almost identical, with houses, rivers, and hills, but facing downward. In this upside-down version of the land of the living, people slept by day and came out at night.

Although the gloomy regions beneath the Earth were believed to be places of the dead, where their souls wandered for eternity, they were also a source of life and wealth. The soil nourished the crops and plants on which people's lives depended. From the rocks of the Earth came gold and precious stones, so valuable that they were often seen as gifts from the gods.

Thankful for these gifts, people often saw the Earth as a place of creation. Many myths tell how people were made from mud and clay, while the Incas of Peru believed that their ancestors came from beneath the Earth. These were three brothers and three sisters, who appeared in three sacred caves near the city of Cuzco. They were dressed in the finest woolen clothes and wore earrings of pure gold to show their royal status.

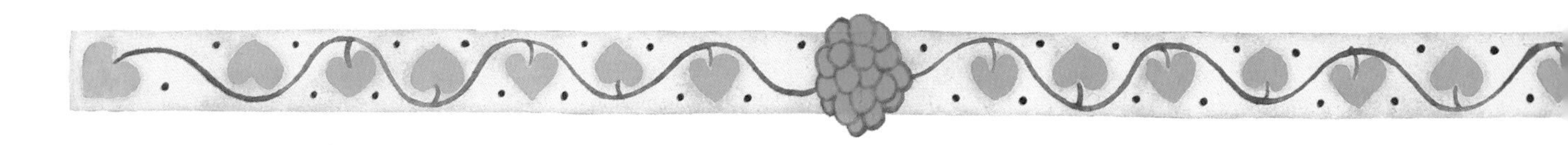

Into the Underworld

In the myths of ancient Greece and Rome, the Underworld — Hades — was the place of the dead. The Roman poet Virgil wrote that the entrance to the Underworld lay deep beneath Lake Avernus, near Naples, Italy. The lake's landscape is volcanic, with pools of mud bubbling from the ground and an eerie stink of sulfur.

Telling the Future

Near Lake Avernus, the hillside is pitted with caves believed to lead to Hades. The most famous is the Cave of the Cumaean Sibyl, an ancient Greek prophetess who gave advice and told the future. It was the Sibyl who led Aeneas, the legendary Roman hero, to the kingdom of the dead — and back again.

Journey to Hades

However, for most Romans there was no returning from Hades. The souls of the dead were rowed across the mythical River Styx to Hades. A coin was placed in the dead person's mouth to pay the ferryman. With its connections to death and decay, Hades also gave the Romans an explanation for the dark months of winter.

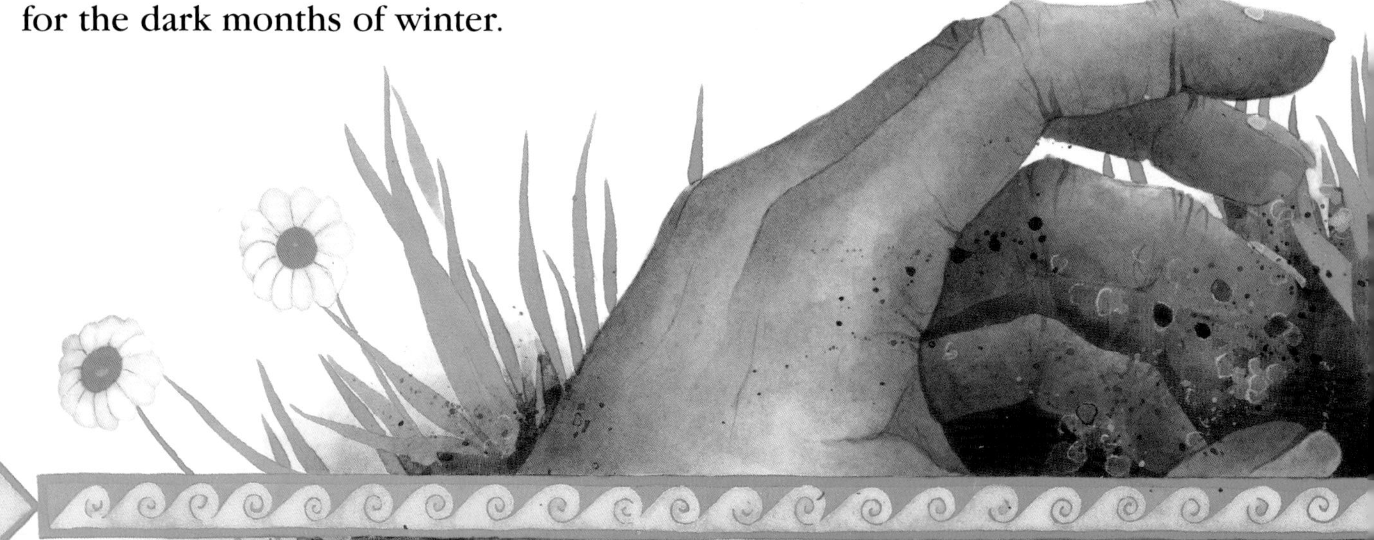

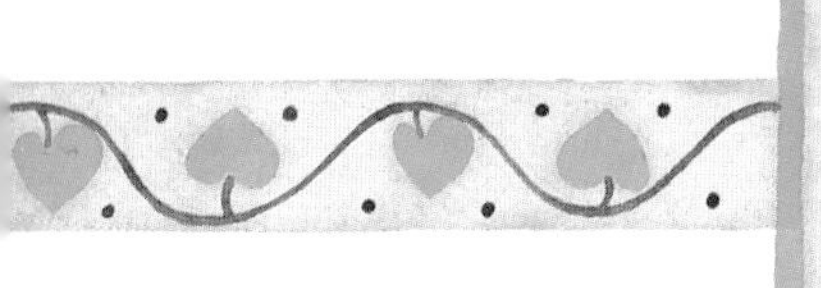

The Goddess and the Pomegranate

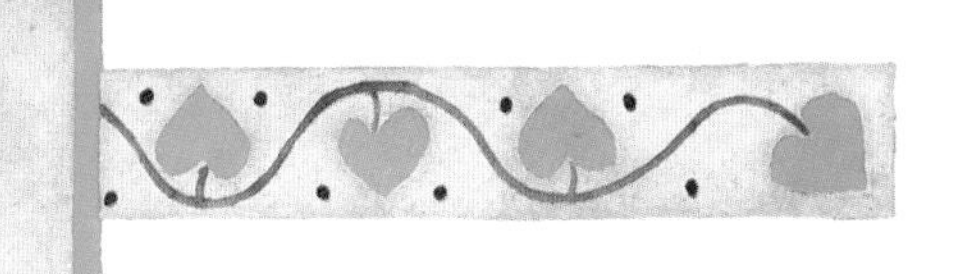

This Roman myth tells how Proserpina was kidnapped by Dis, king of the Underworld. The story is based on the Greek myth of Persephone (Proserpina) and Hades (Dis).

The goddess Ceres, protector of plants and harvests, had a beautiful daughter called Proserpina. One day, Proserpina was picking flowers in the woods, when she suddenly noticed a bright narcissus, a rich, deep yellow like gold.

"I'll pick it for my mother," Proserpina decided. But as she bent down to pluck the flower, a great gaping hole opened up in the ground and from it emerged a terrible figure — Dis, king of the Underworld. He grabbed hold of Proserpina and dragged her down into the depths of the Earth.

Proserpina called out to Jupiter to help her, but the king of the gods was too far away and could not hear her cries. Her mother, Ceres, did hear her voice. For nine days and nights she wandered Earth, without food or sleep, trying to find her daughter. On the tenth day, she met the all-seeing Sun god, who told her what had happened.

"Jupiter is the one you should blame," he told her. "For it was he who gave Dis permission to take Proserpina as his wife."

Ceres was so filled with grief at the loss of her daughter, and so furious at Jupiter's treachery, that she neglected the plants she usually cared for, so that harvests would fail and everything on Earth would wither and die. Then, disguised as a frail old woman, she began her long, long search to find her daughter. Meanwhile, the gods were worried. Unless the crops grew, the whole world would starve. So one by one, they went to Ceres and begged her to change her mind.

"No," replied Ceres. "I will not let the crops grow until I see my daughter again."

There was nothing else to do — Jupiter had to give in. He sent his messenger, Mercury, to the Underworld to bring Proserpina back.

"But," he warned. "If she has eaten anything from Dis's garden, she is doomed to stay in Hades forever."

Proserpina returned from the Underworld, and her mother was filled with joy. But when Ceres asked Proserpina if she had eaten Dis's food, her joy quickly turned to sorrow.

"I only ate some pomegranate," Proserpina said. "Surely that can't do any harm."

But it could. For pomegranates were a symbol of marriage. Ceres begged Jupiter not to take her daughter away again. And Jupiter, mindful that the Earth was still dying, acted kindly.

"You will spend part of the year in Hades with Dis," he decreed. "And the rest on Earth with your mother."

And this is why the Earth has seasons. While Proserpina stays in the Underworld with her husband, Ceres mourns, the plants and crops die, and the world has winter. When Proserpina returns to Earth, Ceres is happy again. The plants begin to bloom and blossom again, and it is spring.

Secret Caves

Carved out from hillsides by water and weathering, caves provided early people with their homes and perhaps their first temples, where they painted magical pictures of the animals they hunted. Caves could be secret and mysterious, too. Some were said to shelter mythical beasts. Others were secret hiding places away from the rest of the world — refuges for hermits or even the king of the gods.

Spirit Shelters

The real-life homes of bears and other animals, caves were often thought to house mythical creatures, from fire-breathing dragons to mischievous spirits. The Native Americans of the northwest coast of North America believed that such Earth spirits lived in caves in the cliffs that lined the seashore. From time to time, the spirits crept out to steal fishing gear or to play other tricks on the fishermen.

Hermit's Home

Often in remote places, some caves provided shelter for holy men and hermits. In the 2nd century A.D., the Christian saint Paul the Hermit spent most of his life living in a cave in the Egyptian desert. Here he could pray and meditate, hidden from his Roman persecutors. Each day a raven brought him half a loaf of bread to eat. When another hermit came to visit, the raven brought a whole loaf!

Birthplace of Zeus

At 8058 feet (2,456 m), Mount Ida, the "high one," is the highest mountain on the island of Crete, in the Mediterranean. In its side is a sacred cave that the ancient Greeks believed to be the birthplace of Zeus, king of the gods. Many pilgrims visited the cave, among them Pythagoras, the famous mathematician. Archaeologists have found many ancient offerings there, including statues and cauldrons, a bronze shield and a drum.

This statue of Zeus dates from the 4th century BC.

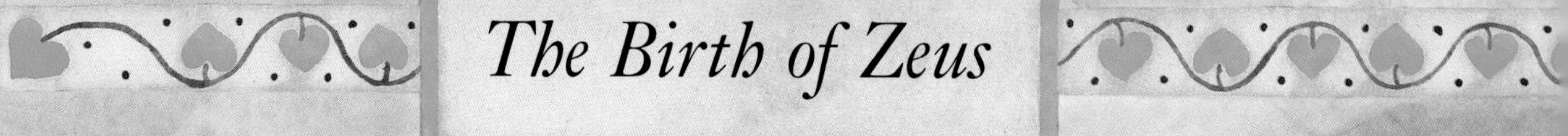

The Birth of Zeus

This is the story of how Zeus was born in a cave on Mount Ida.

✦

Long ago a mighty giant, Kronos, king of the Titans, ruled the Earth. He married the goddess Rhea and had three daughters and three sons. But Kronos had been given a terrible warning — one of his children would one day destroy him and seize his throne. To make sure that this never happened, he swallowed each baby as it was born.

Rhea, his wife, was filled with grief and despair. How could she save her children? When the time drew near for the birth of her sixth child, she fled to Crete and hid in a deep cave on the mountainside. There the baby was born. Rhea called him Zeus.

But the danger was not yet over. Leaving her son for safekeeping with two gentle nymphs, Rhea returned to Kronos. "Give me the child," demanded her husband. And Rhea wrapped a large stone in a blanket, to look like a baby, and gave it to Kronos, who happily swallowed it whole.

Thanks to his mother's quick thinking, Zeus grew up without his father knowing anything of his existence.

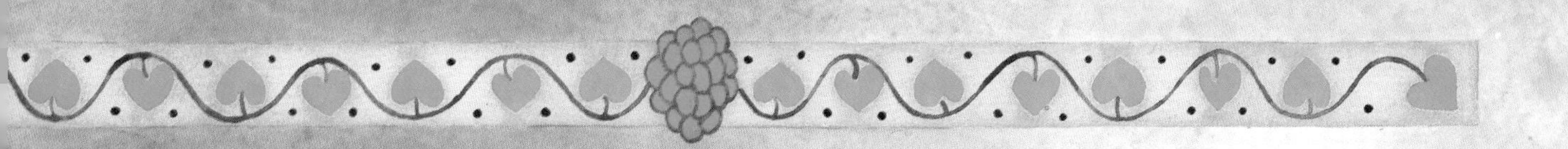

The nymphs brought up the young god with great love and care. They rocked him in a golden cradle and gave him goat's milk to drink. Spirits called the Curetes stood guard outside the cave, clashing their weapons whenever the baby cried, to hide the sound from his hot-tempered father. As soon as Zeus grew up, he vowed to punish his wicked father. But he could not decide what best to do, so he asked Metis, the beautiful goddess of wisdom, for help. Metis had a clever plan. She made her way to Kronos's palace and slipped a powerful potion into his drink — and Kronos began to cough. He coughed and coughed until he brought up all of Zeus's brothers and sisters, swallowed so many years before. He even coughed up Rhea's great stone. Now mighty Zeus appeared and banished his father to the ends of the Earth, where a hundred-armed giant stood guard. Then Zeus married Metis and became king of the gods, ruling the Earth in his father's place, with his brothers and sisters beside him.

Fountains of Fire

Deep beneath the Earth lies a layer of red-hot liquid rock. When gas builds up underground, it pushes the rock upward, until it bursts to the surface in a cloud of ash, smoke, and steam — a volcano is erupting. The molten rock floods down the volcano's sides in glowing rivers. No one can be sure how far they will stretch.

Workshop of the Gods

An erupting volcano is one of the most awesome and terrifying sights in nature. Before their true cause was understood, most people believed that the explosions were the work of angry gods. Indeed, volcanoes are named after the Roman god Vulcan, blacksmith to the gods and god of fire. He had his workshop inside the Solfatara volcano, near Naples, Italy. The rumbling sound that came from the volcano was said to be Vulcan hammering away at his forge.

Hawaiian Volcanoes

The islands that make up Hawaii are the tops of huge underwater volcanoes that have bubbled up over millions of years and are still bubbling! Between 1843 and 1984, the volcano Mauna Loa erupted about once every four and a half years. But Kilauea is the world's most active volcano. It has erupted nonstop since 1983.

A Goddess's Anger

According to legend, Kilauea is the home of the fiery goddess Pele, the creator of Hawaii. She is said to live deep inside the volcano's crater. When she is angry, she stamps her feet and makes the volcano erupt. In ancient times, people built temples to Pele on the volcano's slopes and threw offerings of food, drink, and flowers into the crater to keep her happy. Just before an eruption, Pele was said to appear, disguised as an old woman. Other people claimed to see her face in the clouds.

A wooden carving of Pele

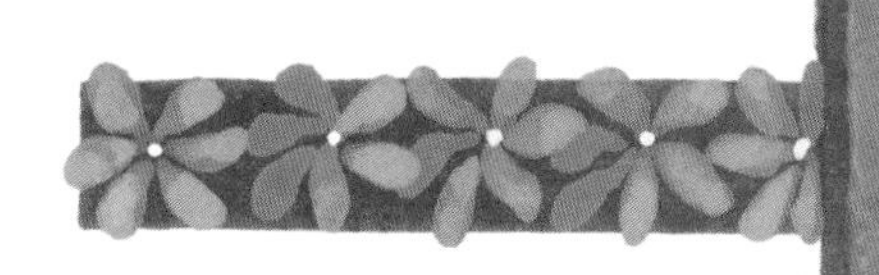

How Pele Found a Home

This is the story of how the goddess Pele came to live in Kilauea.

✦

Some people say that the goddess Pele came from the island of Tahiti. She longed to travel, and one day, tucking her sister under her arm, she set off in a great carved canoe. With her went a whole host of gods, to steer the canoe over the ocean.

So Pele traveled from island to island, looking for a place to live. Wherever she landed, lightning flashed, thunder rumbled and raged, and red-hot tongues of lava sped across the ground. But Pele could not find a home. Whenever she dug a hole in the ground, the sea rolled in and drove her away. At last she dug a deep, deep hole in the great volcano of Kilauea. It was solid rock all the way down. And there Pele stayed.

Although Pele was beautiful, she was very hot tempered. She only had to stamp her foot and a volcano erupted somewhere nearby. With one terrible glance, she could turn people to stone. Woe betide anyone who made her angry. This did not stop Kamapua'a, the Pig-man, from falling madly in love with her. Kamapua'a had the power to turn himself into a pig. When he appeared as a human, he wore a long cloak to hide the bristles on his back. Pele could not stand the sight of him.

"Go away and leave me alone," she shrieked. "I don't want to marry a pig. Even if you were the last pig alive."

Their quarrels grew worse and worse. Pele covered the Pig-man in showers of flames and nearly destroyed him. He, in turn, put out her fires with fog and rain, and overran her home with pigs. Soon all the fires on Hawaii went out, and the islands were plunged into darkness. It was time for the gods to act, and quickly.

To keep Pele and the Pig-man apart, the gods divided the islands of Hawaii between the two of them. The places where the lava flowed belonged to Pele; the places where it always rained to the Pig-man. And so fire returned to the island, bursting up from the ground and lighting up the sky whenever the goddess is angry.

The Shaking Earth

The Earth's crust is not a solid sphere of rock. It is cracked into gigantic pieces called plates, which rest on the layer of softer rock beneath. The plates are constantly pushed and jostled together, putting the rocks of the crust under great strain. All of a sudden, the ground can give way with a mighty crack. This is an earthquake. Some earthquakes are hardly felt. Others can reduce whole cities to rubble.

Early Beliefs

The mysterious quaking of the Earth baffled ancient people. Could it be caused by a gigantic animal? Some blamed the burrowing of a giant mole. Others blamed fish. One African myth tells of a huge fish that carried a stone on its back. The stone supported a cow that balanced the world on its horns. Earthquakes happened as the cow juggled the world from horn to horn. The Japanese, too, believed a fish was the cause of earthquakes.

How a Giant Catfish Shook the World

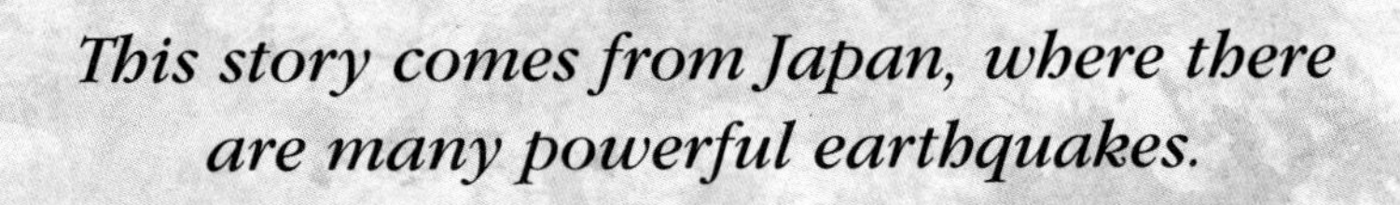

This story comes from Japan, where there are many powerful earthquakes.

✦

People dreaded the time of year when the gods left Earth to return to heaven, for then the ground began to tremble and shake. Great cracks opened up beneath people's feet, swallowing up houses, fields, and villages.

"We must ask the gods to save us," the people said. They prayed to the gods and left gifts of flowers at their shrines, until the gods agreed to help. But first they had to find out what was making the Earth shake.

The gods searched high and low — on the tops of mountains, under the sea, and even in heaven itself. Then they looked deep beneath the Earth, where they found a gigantic catfish. Normally the fish slept, lying quite still in its underground home. But sometimes it grew restless. It wriggled and writhed and flapped its tail, and this is what made the ground shake. The gods picked up an enormous stone and placed it right on top of the catfish so that it could not move an inch. "And that's the end of that," they said happily.

But it wasn't, quite. Every now and again, the catfish gives a great wriggle and pushes the stone away. And then the ground trembles and shakes, and great cracks open up beneath people's feet, until the gods replace the stone.

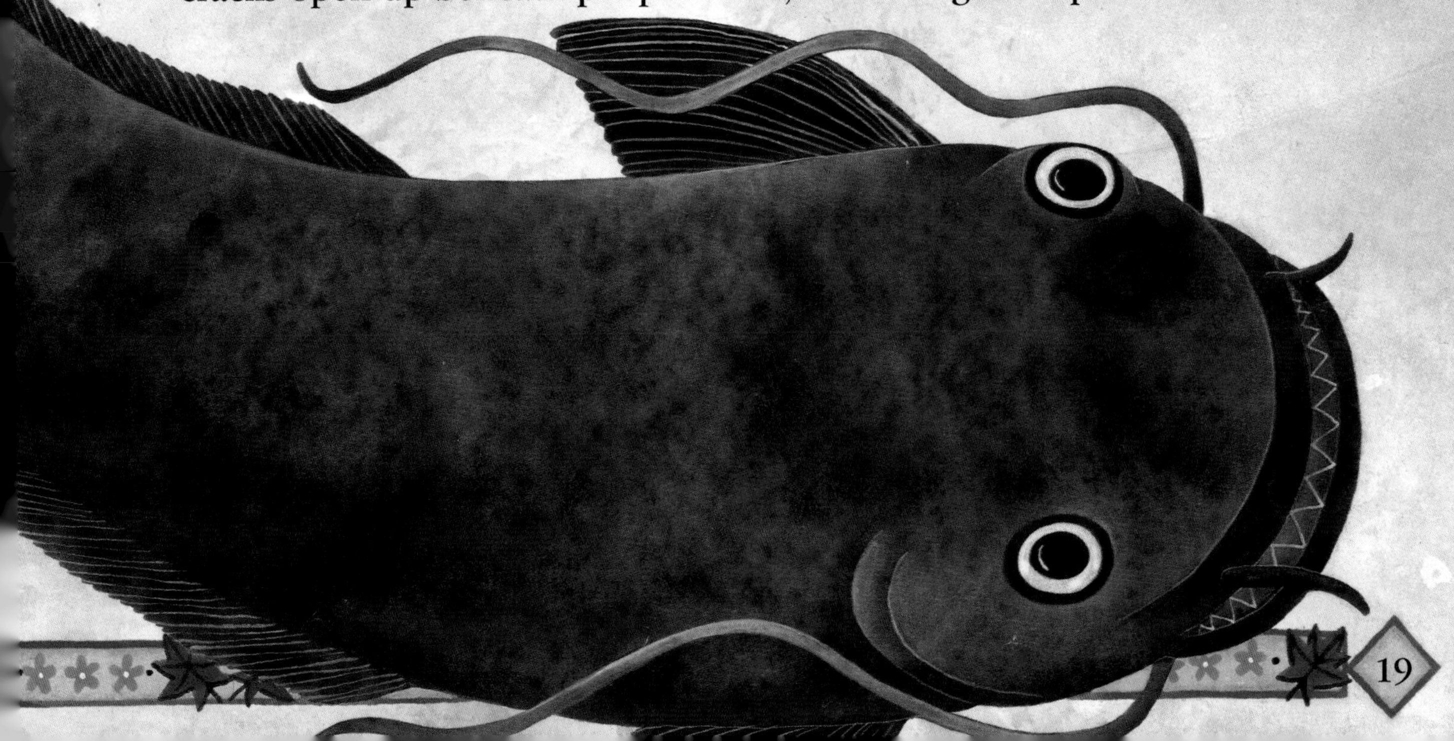

Springs of Life

Beneath the Earth's hills and plains lies a layer of water, formed by rain that has seeped underground. In some places, the water bubbles to the surface, making lakes, wells, gushing springs, and welcome oases in the desert. Spring water is usually fresh and pure because it is filtered by the rocks it passes through or contains chemicals known for their medical properties. No wonder that ancient people thought these waters had magical powers of healing.

A Source of Wisdom

Because springs come from beneath the Earth, their source was believed to be the Underworld, and their waters were thought to have mysterious properties. Some springs were said to have the power to grant wisdom and make wishes come true, and bathing in others could make you immortal.

Center of the World

The Castalian Spring at Delphi, Greece, runs from a deep crack in the cliffs. Greek myth says that it was created by the hooves of the winged horse Pegasus and that it lies at the center of the world. In ancient times, visitors bathed in the spring before consulting the oracle — a priestess who spoke with the voice of Apollo, the sun god. Apollo was also god of poetry, and the spring became famous for inspiring poets to produce their greatest works.

River Beginnings

In Irish legend, the Well of Seghais was also a source of inspiration. This sacred spring fed the two great rivers of Ireland, the Shannon and the Boyne. The spring was said to be home to a magical salmon of great knowledge. Anyone skillful enough to catch and eat the salmon would gain its wisdom for himself. Legends such as this probably come from the Celts, ancestors of the Irish, and have been passed down by word of mouth over many centuries.

Finn and the Salmon of Knowledge

This is the story of how the great Irish hero Finn MacCool gained his supernatural knowledge.

When the great warrior and giant Finn MacCool was a young boy, he went to visit a poet who lived by the Well of Seghais. The poet, whose skill and fame had reached far and wide, had promised to teach the young Finn all he knew about wisdom, knowledge, and the world. For seven long years, the poet himself had waited patiently by the well, hoping against hope that he would see the magical salmon that lived in its waters. For it had been foretold that whoever could catch and eat the salmon would gain the gift of prophecy and great knowledge beyond anything known to ordinary folk. Around the well grew nine magical hazel trees, delicate and green. The salmon had gained its mysterious powers by eating the hazelnuts that fell from the trees into the magical waters of the well. And so it became the Salmon of Knowledge.

One day, the poet's wish came true. At long, long last, he caught the salmon and gave it to Finn to cook.

"But don't eat any of it, not even a bit," he warned Finn.

Finn did exactly as the poet said. He cooked the salmon and brought it to his teacher.

"I did as you said," Finn told the poet. "But the fish was hot when it came out of the pot, and I burned my thumb on its skin. Then I put my thumb in my mouth to cool."

The poet gave Finn a very long look, filled with a mixture of anger and awe.

"Then you already have the knowledge," he said, in a somber voice. "And you must eat the fish, not I."

So it was that Finn ate the Salmon of Knowledge. And from that day on, whenever he needed to call upon its magic powers, he only had to suck his thumb, and what he did not know was immediately revealed to him.

Riches from the Earth

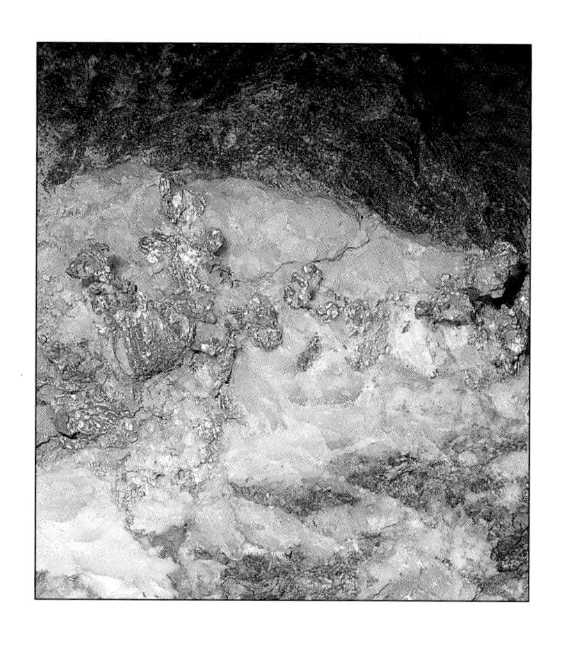

For thousands of years, people have mined the riches that lie beneath the Earth — metals, minerals, and glittering gemstones. These jewels were widely believed to have magical powers. Emeralds, for example, were said to be able to cure diseases and calm the stormy seas. Most precious of all was gold, worshiped as a symbol of wisdom and immortality and treasured by kings and queens.

God of Iron

For some cultures iron was even more important than gold. It was essential for making tools and weapons and was often worshiped as a god. In African legend, the god of iron was called Ogun. He was a hunter. When the Great God decided to live on Earth, it was Ogun who cleared the forest for him with his iron ax. He was given a kingdom as a reward. But Ogun did not want to rule. He preferred to live alone, on a hilltop, where he could watch over the land and spy out his prey.

This African man is carrying an iron weapon in a festival to celebrate Ogun.

Viking Blacksmiths

The Vikings were daring sailors and warriors from Scandinavia. They also closely associated metals and gems with their gods. A Viking warrior's most prized possession was his sword, and the blacksmith who made it was greatly admired. The Viking god of blacksmiths was called Weland. It was said that if you left a broken horseshoe and a coin near Weland's home, the horseshoe would be magically mended when you returned.

The Dwarfs' Treasure

In Viking myth, the rightful owners of the gold and gemstones buried underground were the Dwarfs. Expert craftsmen, they made exquisite gifts for the gods and goddesses. Among these were the locks of golden hair made for the goddess Sif, which grew just like real hair, and a magic golden boat for the god Frey. It was big enough to hold all the gods but could be folded up very small when not in use. Many wanted the Dwarfs' riches, and the god Loki often made mischief for them in order to steal their treasure.

Andvari's Gold

This is the story of how Andvari the Dwarf's treasure was stolen by the god of mischief, Loki.

✦

Many years ago, three gods were on a visit to the land of humankind, dressed as ordinary men — Odin, father of the gods; Loki, god of mischief-making; and Honir, the great warrior. As they strolled along a riverbank, they spied an otter catching a salmon.

"We'll have that salmon for our supper," said Loki. He threw a stone, which killed the otter. The fish was theirs. Loki cooked the salmon for his friends. Then he cut off the otter's fine skin to keep. That night the three men asked for shelter at a nearby farm. The farmer, Hreidmar, was gruff and rude, but he took the men in and soon they fell asleep. As soon as their eyes closed, Hreidmar's sons seized their guests and tied them up.

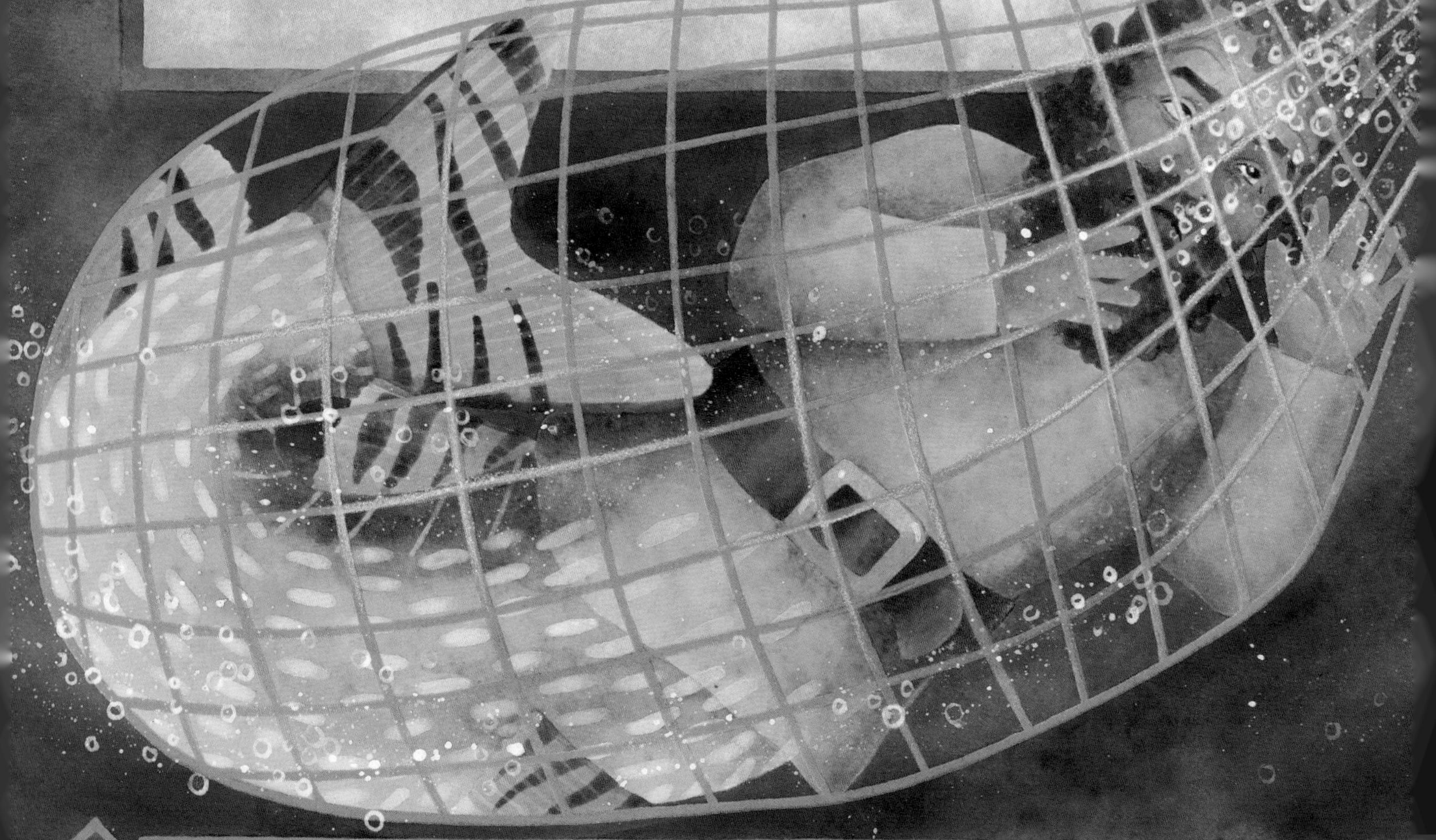

"I should kill you here and now," growled Hreidmar. "For that otter skin you carry belonged to my son. Every day, I turned him into an otter to go fishing. And you have killed him."

The gods were astonished. They offered to repay Hreidmar for what they had done, if only he would spare their lives. "Cover the otter skin with gold," he told them. "Then I will set you free."

Loki was sent to fetch the gold. He dived to the bottom of the sea to borrow the sea goddess's magic net, which caught whatever you told it to. Then he found a deep, dark underground pool. He cast the net and caught a huge pike, which immediately turned into Andvari the Dwarf, who could take the shape of a fish. Andvari was famous for his fabulous store of gold, which he kept well hidden from prying eyes. He was sure to have gold aplenty for covering an otter skin.

"Unless you give me your gold," Loki threatened, "I shall keep you as my prisoner forever. Which is it to be?"

Andvari had no choice. Fearing for his life, he led Loki to his secret store and helped him fill his sack with gold. Just as he was about to leave, Loki spotted Andvari's golden ring and demanded it from him.

"Take it," said Andvari. "But be warned. Whoever wears that ring is cursed. The ring will destroy its wearer."

But Loki was not listening. He took the gold and returned to Hreidmar, who, true to his word, set Odin and Honir free. The otter skin was covered in gold, from head to tail, apart from one last whisker. The only gold left was Andvari's ring, which Loki handed over, making no mention of the curse that went with it.

It was not long before Hreidmar's family began to argue over the gold. Greedy to get his hands on it, Hreidmar's son Fafnir murdered his father and ran away with the gold so he did not have to share it with his brother. He buried it away deep underground and turned into a dragon to guard it. Was the curse of the ring at work?

It seemed it was, for Hreidmar's sons and grandsons fought and killed each other for the gold over the years to come, until none of the men in the family was left alive. As the Dwarf himself had foretold, Andvari's gold brought nothing but misery to those who owned it.

Notes and Explanations

Who's Who

ANCIENT GREEKS: The people who lived in Greece from about 2000 B.C.. They worshiped many gods and goddesses who controlled the natural world and daily life.

ANCIENT ROMANS: The people who lived in Italy from about 750 B.C. to A.D. 476, named after the ancient city of Rome, founded in about 753 B.C.. By 100 B.C., the Romans ruled over a huge empire centered around the Mediterranean Sea.

CELTS: A warlike people who first lived in Austria and France from about 600 B.C.. Celtic tribes gradually spread across southern and western Europe, traveling over the sea to Britain and Ireland. The Celts were famous for their beautiful jewelery, music, and poetry, which told the stories of their gods and heroes.

CHRISTIANS: Followers of Jesus Christ, a teacher and preacher who lived 2,000 years ago in Palestine. Christians believe he is the Son of God. Today, there are about 200 million Christians, all over the world.

INCAS: The people who lived in the Andes mountains of South America. At the beginning of the 16th century, they ruled over an empire that covered much of modern-day Peru, Ecuador, Bolivia, and Chile. By 1533, they themselves had been conquered by Spanish invaders.

PYTHAGORAS: A Greek mathematician and philosopher who was born about 580 B.C.. Pythagoras is most famous for his work in geometry. Many of his ideas are still used today, including his famous theorem on right-angled triangles.

VIKINGS: The Vikings were a seafaring people from Scandinavia who raided and conquered many parts of northern Europe between the 8th and 11th centuries A.D..

VIRGIL: A Roman poet who lived from 70-19 B.C.. His most famous work is called *The Aeneid*, a long poem that tells the story of the origin and growth of the Roman Empire. Its hero is a prince called Aeneas, who was the ancestor of Romulus and Remus, the legendary founders of Rome.

What's What

Strictly speaking, fables, legends, and myths are all slightly different. But the three terms are often used to mean the same thing — a symbolic story or a story with a message.

FABLE: A short story, not based on fact, that often has animals as its central characters and a strong moral lesson to teach.

LEGEND: An ancient, traditional story based on supposed historical figures or events. Many legends are based on myths.

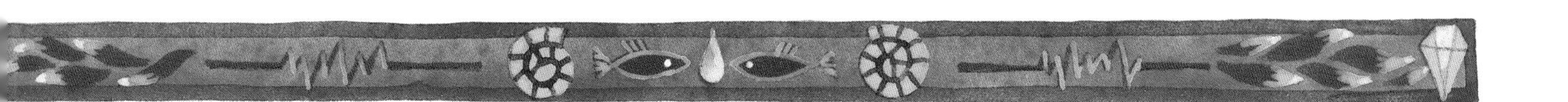

MYTH: A story that is not based in historical fact but that uses supernatural characters to explain natural phenomena, such as the weather, night and day, and so on.

IMMORTALITY: Immortality means living for ever, never dying or aging. In many religions, the gods were believed to be immortal.

ORACLE: A sacred place where people went to consult a god or goddess. Oracle also means the priest or priestess who spoke to people on the god's behalf.

PILGRIMS: People who make a special journey, or pilgrimage, to a sacred place as an act of religious faith and devotion. Sacred places include natural sites, such as rivers and mountains, and places connected with events in the history of a religion.

PLATES: The Earth's crust is not a solid sheet of rock. It is cracked into huge pieces called plates. These float on a layer of soft rock beneath. As the plates collide and jerk away from each other, they cause earthquakes and volcanoes.

SOUL: The inner part or spirit of a person or animal, as opposed to its physical body. Many people believe the soul gives life to the body, and, unlike the body, it never dies.

SULFUR: Sulfur is a strong-smelling gas that is given off when a volcano erupts. It smells like rotten eggs.

Where's Where

The map below shows where in the world the places named in this book are found.

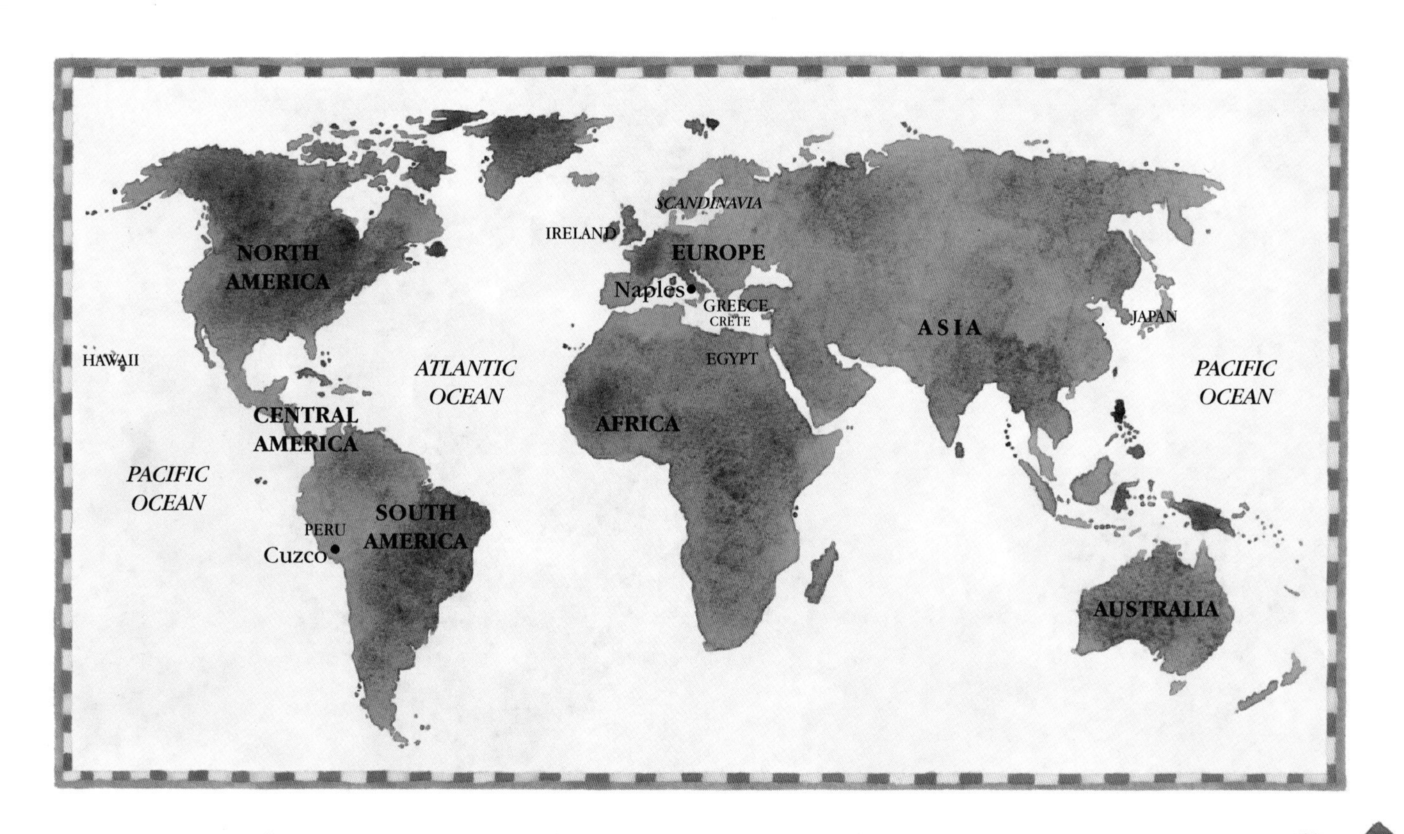

Index